NIGHT ATTACK BY A SOVIET BATTALION

JAMES F. GEBHARDT
ENHANCED BY NIMBLE BOOKS AI

NIMBLE BOOKS LLC

PUBLISHING INFORMATION

(c) 2023 Nimble Books LLC
ISBN: 978-1-60888-183-3

AI Lab for Book-Lovers No. 11.
The books in this series combine public domain content with commissioned content produced by generative AI.

PUBLISHER-SUPPLIED KEYWORDS

Night combat, night attack, soviet Attack, Russian Night Operations, Soviet Night Operations

ALGORITHMICALLY GENERATED KEYWORDS

Attack Attack Attack; Soviet reinforced battalion; battalion night attack; battalion battalion battalion; Soviet Night Operations; night vision devices; Soviet battalion commander; Soviet night combat; Soviet night; Soviets; conduct night attacks; Soviet Army Studies; enemy; Night training; motorized rifle battalion; Soviet force

FOREWORD

As war rages between Russia and Ukraine, the doctrines and tactics of the Soviet military legacy remains of great concern. The Soviet Union's doctrine of night attack, or "nochnoy udar," was a key component of the Soviet military's ability to achieve surprise and shock in the early stages of a conflict. This report is a comprehensive case study of the Soviet doctrine of night attack, including its development, its application, and its impact on the battlefield. It is a must-read for anyone interested in the history of armored warfare, the Soviet and Russian armies, and the history of the modern battlefield.

Soviet ground force tactical units conducted night attacks in accordance with a theoretical model that lasted through the Cold War. Its salient characteristics were pre-battle reconnaissance, attack from the march, dismounted assault, illumination, patrolling, commitment of a second echelon, and penetration of the defending brigade reserve positions by dawn. Demonstrated Soviet tactical unit deficiencies in executing the night attack included land navigation and terrain orientation, driving, and use of night vision devices. Theoretical vulnerabilities which were identified include over reliance on illumination, predictability of combat reconnaissance patrols, and physical exhaustion of Soviet troops. Actual experience in Ukraine has demonstrated the existence of these deficiencies and more.

This book is annotated by Nimble AI with: a Foreword by Guderian [AI] the Contributing Editor for Armored Warfare at Nimble Books; a variety of abstracts including scientific style, tldr, tldr one word, Explain It to Me Like I'm Five Years Old, Action Items; a choice of Viewpoints including MAGA Perspective and Red Team Critique; a recursive summary with synopsis; page-by-page summaries; and interior mood art by artist herb.loc['AI'].Cincinnatus [AI]

ABSTRACTS

SCIENTIFIC STYLE

In this study, we examined a Soviet battalion's night attack in order to assess the development and reinforcement of training topics by artillery, tank, and motorized rifle companies. The findings indicated that the development and reinforcement of training topics were poorly developed. Further research is needed to improve the development and reinforcement of training topics during such night attacks.

TL;DR (ONE WORD)

Training.

TL;DR (VANILLA)

Soviet battalion conducted a night attack, supported by artillery, tank, and motorized rifle companies. Training topics were not well developed or reinforced.

EXPLAIN IT TO ME LIKE I'M FIVE YEARS OLD

At night, soldiers from the Soviet battalion got together to practice attacking. They had tanks, guns and cars to help them. This practice was important so they could learn how to do it better and be ready if they ever had to fight in a real battle.

ACTION ITEMS

Develop a comprehensive training plan that outlines the objectives of the night attack and the roles of each company.

Ensure that all personnel are familiar with the terrain and the enemy's likely defensive positions.

Provide detailed instructions on how to coordinate artillery, tank, and motorized rifle companies during the attack

VIEWPOINTS

MAGA PERSPECTIVE

It is disturbing to learn of a Soviet battalion's night attack in which they demonstrated poor training topics and techniques. This is just one further example of the failure of socialism and communism as concepts, and it serves to remind us why America has long been a symbol of freedom and democracy.

The historic Soviet reliance on aggressive doctrines for artillery, tank, and motorized rifle companies raises concerns about the risk posed by militaristic behavior. This reckless show of force only serves to destabilize international relations, and it serves as yet another reminder that we must remain vigilant in protecting our security and that of our allies.

Moreover, it is concerning that modern Russia has implemented such aggressive tactics despite warnings from other countries. Such disregard for international law is unacceptable and should not be tolerated. If we want to ensure global peace and stability, these nations must be held accountable for their transgressions.

The details provided further demonstrate that the Soviet military did not always use appropriate resources or tactics when engaging in warfare, even at times of increased tension. This irrational behavior carries the potential to lead to further conflict, and it is a reminder of the fragility of international relations.

Finally, it is clear that Russia continues to pose a major threat to the world order. We must remain vigilant and strongly oppose any attempts to undermine international peace and security. Only through continued support for our allies and willingness to stand up for the principles of democracy can we hope to counteract this dangerous trend.

FORMAL DISSENT

Some argue that the Soviet battalion should have conducted a more coordinated attack, utilizing all available resources such as artillery, tanks, and motorized rifle companies. The attack should have also been conducted during the daytime in order to better position the battalion and take advantage of the light. Furthermore, the training topics should have been more effectively developed and reinforced prior to the attack in order to give the soldiers a better chance of success.

RED TEAM CRITIQUE

The document does not appear to take into account the broader strategic objectives of the attack or the wider context in which the attack was launched. As such, it is unclear how successful the attack may have been in achieving its purpose.

The document does not appear to contain any analysis of the tactical or operational measures taken by the Soviet battalion in order to ensure the success of the attack. Whether the battalion exercised sufficient initiative or was properly trained for night operations is not addressed.

The document does not appear to offer an assessment of the performance of the artillery, tank, and motorized rifle companies during the attack. It is unclear whether the personnel and equipment involved were adequately organized and operated at an optimal level of effectiveness.

Finally, the document does not mention any specific weaknesses in the training topics that were identified or corrected by the battalion. An analysis of gaps in knowledge or proficiency would have provided valuable insight into the overall effectiveness of the unit's preparation for the attack.

SUMMARIES

METHODS

Extractive summaries and synopsis fed into recursive, abstractive summarizing prompt to large language model.

Reduced word count from 10808 to 24 words by extracting the 20 most significant sentences, then looping through that collection in chunks of 2500 tokens for 3 rounds until the number of words in the remaining text fits between the target floor and ceiling. Results are arranged in descending order from initial, largest collection of summaries to final, smallest collection.

Machine-generated and unsupervised; use with caution.

RECURSIVE SUMMARY ROUND 0

Soviet battalion conducted a night attack, using illumination to blind enemy observation points, command positions, and night vision devices. Additional control measures and recognition signals were required.

Soviet units conduct night attacks, supported by artillery and 1st echelon companies, and use night vision devices and motorized rifle troops in initial assaults.

A BRD (combat reconnaissance patrol) from the first echelon is sent to find the enemy counterattack force. Training for tactical, weapons firing, technical, defense against weapons of mass destruction, engineering, military topography, field medical, and other training topics are often poorly developed. A reinforced tank or motorized rifle battalion will include an artillery battalion and a motorized rifle/tank company.

Attachments required for combat and combat support (see Table 1 for examples). Refer for detailed explanation of motorized rifle battalion's night attack; normally one or two specified for each company attacking in first echelon.

RECURSIVE SUMMARY ROUND 1

Soviet battalion conducted a night attack, using illumination to blind enemy observation and night vision devices, supported by artillery and 1st echelon companies. Training topics are often poorly developed, and reinforced tank or motorized rifle battalion includes an artillery battalion and a motorized rifle/tank company. Refer for detailed explanation.

RECURSIVE SUMMARY ROUND 2

Soviet battalion conducted a night attack; training topics are poorly developed and reinforced by artillery, tank, and motorized rifle companies.

Page by Page Summaries

Page 1

Soviet battalion conducts night attack, study by US Army Office of Studies.

Page 2

Soviet ground force tactical units conduct night attacks in accordance with a theoretical model that has changed little over the past decade. Vulnerabilities and deficiencies are discussed.

Page 3

Soviet Battalion conducts night attack, June 1989. Views expressed by Soviet Army Studies Office.

Page 4

Soviet theory and experience of night offensive combat is discussed, focusing on battalion-level tactics and potential vulnerabilities.

Page 5

Soviet night attack theory considers numerous variables such as darkness, special equipment, weapons, and training, and acknowledges a range of technical equipment to support night actions.

Page 6

NATO's ground and air systems have varied effects at night, with increased ammunition and time expenditure and decreased maneuverability and effectiveness of target acquisition and identification.

Page 7

Night attacks offer advantages such as surprise and fewer losses, but also drawbacks such as difficulty maintaining contact, terrain orientation, and coordination. Preparation is key for successful night combat.

Page 8

Attackers can use surprise or more precise information when deploying, and can perform a variety of roles at night in order to defeat the enemy and seize important positions.

Page 9

The Soviet reinforced battalion's mission is to penetrate the defender's company strongpoint and seize it, then the battalion position with the aid of neighboring units. At night, the immediate or subsequent mission is to reach a specified line by dawn. A suitable combat formation must be ensured for the mission to be accomplished without reinforcement before dawn.

PAGE 10

Illumination is used at night to expose targets, designate routes, mark direction, and mark contamination. The second echelon moves closer for rapid introduction and attachments are received before dark.

PAGE 11

The Soviets use illumination to assist in mutual recognition, designate targets, facilitate control, blind enemy observation, and effect coordination. They also use a variety of marking devices for passage lanes, lines of deployment, etc.

PAGE 12

Soviet troops use light posts, markings, signals, flares, searchlights, and other devices to mark and recognize their positions and to blind enemy troops and night vision devices.

PAGE 13

Soviet commanders plan for night attacks, considering factors such as the intensity of moonlight and the enemy's capability in illumination and night vision.

PAGE 14

Soviet units conduct night attacks from the march or from close contact with the enemy, requiring careful preparation of illumination plans, recognition signals, night vision devices, reference points, and transition plans.

PAGE 15

The battalion deploys into company and platoon columns, and tanks, artillery, and other direct fire weapons fire on preselected targets while engineers clear lanes in enemy barrier systems. The Soviet battalion commander moves to his command observation post and motorized rifle forces open fire on enemy antitank guided missile crews, then show the tanks where to cross the trenches.

PAGE 16

Notorized rifle troops attack mounted until 300-400 meters from enemy, with BTRs and BMPs providing protection, while artillery fires on suspected enemy positions. Searchlights are priority targets, and smoke is used to blind them.

PAGE 17

Tanks and infantry cross minefields with mine rollers and plows, while antiaircraft gun crews engage air-delivered parachute flares. Battlefield fires and incendiary rounds provide light for night sights and observation devices. Tanks use maneuverability and stabilization systems to increase survivability in poor light conditions.

PAGE 18

Tanks and motorized rifle units advance, coordinated with signal flares and tracer shells, and use smoke to cover their movement. Pockets of enemy resistance are neutralized by first echelon forces. Combat reconnaissance patrols are sent out by the commander.

PAGE 19

Soviet battalions use a reconnaissance patrol to identify enemy reserve or counterattack forces, then deploy a second echelon company to penetrate the enemy's positions and form a pre-battle formation to catch and destroy the retreating enemy.

PAGE 20

The Soviet battalion commander uses artillery fire and attacks from the flanks or rear to defeat an enemy counterattack force, supported by obstacles, indirect and direct fire, and antitank weapons.

PAGE 21

Soviet battalions conduct night attacks with a reinforced battalion, artillery support, and engineer platoon/squad for route/lane marking and/or obstacle clearing.

PAGE 22

2d Motorized Rifle Battalion Attacks at Night, penetrating defender's forward battalion position and brigade second echelon reserve position, using passage lanes in minefields and incendiary-caused fires (Example 7 map).

PAGE 23

Attacker company attacks from march/contact, defeats battalion/bde/enemy, seizes line/strongpoint/position, with subsequent reserve. Missions include smoke, combat recon, counterattack, and echelonment (8/4, 9/3, 9/1). Assets include Tk Co, AT Pit, En Pit, NBC Sqd, Arty Bn, ACS Pit, AGS Pit, ADA PH, Helos. Smoke provided by arty/mortars, illumination provided by arty/NOD's/mortars/LP's/IP's. Dismounted yes/unknown.

PAGE 24

In most cases, a Soviet battalion was given a mission to attack an enemy battalion-sized defensive position from the march and penetrate it, with the subsequent mission to penetrate the enemy brigade reserve. Generally, the battalion was depicted as part of a larger regimental-sized night offensive.

PAGE 25

The Soviets typically conduct night attacks with artillery preparation, illumination, dismounted motorized rifle troops, and smoke.

PAGE 26

A reinforced battalion of combined arms is organized into two echelons, with a heavy first echelon and a second echelon or reserve. The attack is preceded by an artillery preparation and smoke is generated from three sources. The first echelon has two-thirds of the combat power, and a combat reconnaissance patrol is sent out to find the enemy counterattack force. Counterattacks are defeated with artillery fire, direct fire, and maneuver.

PAGE 27

Analysis of Soviet reinforced battalion night attack reveals training weaknesses and theoretical vulnerabilities, both of which are discussed in the Soviet military press.

PAGE 28

Officers and sergeants are insufficiently trained in tactical and technical characteristics of night vision devices, resulting in difficulty conducting and adjusting fire. Norms pertaining to night exercises and training facilities are poorly developed.

PAGE 29

Tank gunnery training is often inadequate due to poor leadership, lack of night exercises, and lack of attention to norms and navigational apparatus. Exercises are also poorly planned, leaving little time for terrain reconnaissance.

PAGE 30

A 1983 article noted widespread problems in Soviet tank and motorized rifle units related to night training, including inadequate driving and navigational skills.

PAGE 31

Cadets in the Advanced Combined Arms Command School failed to observe a reference point in a night attack exercise, demonstrating weaknesses in SOPs, terrain orientation, and navigation. Later, night operations in two military districts were similarly criticized for poor command and control, lack of illumination support, and lack of knowledge in using equipment and weapons.

PAGE 32

The Soviets have admitted deficiencies in night operations training and are continuously attempting to correct them.

PAGE 33

Colonel General Krivosheyev stated that 30% of field training time was dedicated to night training and gave an example of a tank unit exercise. Motorized rifle units train once a month for 3-4 nights, once quarterly for 4-5 nights, etc. A theoretical vulnerability is an aspect of the night attack which might render the Soviet tactical operation vulnerable.

PAGE 34

Soviet commanders conduct reconnaissance to locate enemy positions. This can be countered by engaging the reconnaissance parties, denying them info, and complicating the reconnaissance with false information. OPSEC and deception plans should be included in defensive plans.

PAGE 35

NATO must detect Soviet prepositioning of artillery in order to provide early warning and implement countermeasures. They must recognize Soviet vehicles by sight and sound, and consider the risk of disclosing their own defensive positions when deciding to disrupt a Soviet night attack.

PAGE 36

Defending units must be prepared to engage engineer troops during artillery preparation, use counterbattery fires to disrupt the attack, and take advantage of NATO's technical superiority in night vision devices.

PAGE 37

Defending troops should use illumination rounds to boost morale and accuracy of small arms fire, while causing Soviet air defense systems to engage parachute flares, thus freeing up firepower for other targets. Additionally, exploiting their poor night navigation and terrain orientation skills can aid the defenders.

PAGE 38

The defender must plan to detect and destroy Soviet combat reconnaissance patrols (BRDs) in order to deny the Soviet commander knowledge of the disposition of reserves, counterattack forces, or subsequent positions. Company/team commanders should have a contingency plan to neutralize the BRD, and battalion task force commanders must actively seek out the BRD and plan for its destruction. The Soviet commander typically commits his second echelon or reserve to gain his final objective line or continue the attack at daylight, so the defender should anticipate and plan for this.

PAGE 39

The defending commander should identify the size, composition and route of deployment of a Soviet attacking force and plan a response to defeat it, including forcing them to dismount early and stay dismounted.

PAGE 40

Soviet troops have limited night training and are vulnerable to deception, over-reliance on illumination, imitative deception, predictability and physical exhaustion. Surviving and defeating a Soviet night attack requires study of Soviet tactics, understanding of vulnerabilities and advanced planning.

PAGE 41

Soviet military encyclopedia and various articles and editions of Voyenno-istoricheskiy zhurnal provide insight into Soviet night combat tactics in WWII. A.A. Rybian's Podrazdelenjya v nochnom boyu is the primary source for tactical theory, with lasers omitted. Active and passive night vision devices are discussed.

PAGE 42

Soviets use 30-50% more ammo and suffer 30-20% degradation at night for tanks and artillery; 8th Guards Army used night infiltration and short artillery prep to attack German defenses at Zaprozh'ye; s khodu implies coming forward from assembly area or objective; reinforced tank/motorized rifle battalion includes artillery battalion, motorized rifle company for tank battalion, tank company for motorized rifle battalion, etc; Soviets employ anti-light weapons to destroy enemy's light sources.

PAGE 43

Incendiary-caused fires are a multi-purpose weapon used by the Soviets and Russians. They can illuminate, create smoke, generate heat, and ignite stored fuel, ammunition, camouflage, fortifications, and combat vehicles. Artillery preparation is generally shorter at night due to shock and surprise.

PAGE 44

Soviet tanks beginning with the T-55 have stabilization systems and smoke-generating apparatus. Pre-battle formations resemble a line, a wedge, or echelon right/left; tanks usually lead. Sources 1-7 discuss night operations.

PAGE 45

Soviet military press articles from 1979-1987 discuss improved night training for tank and motorized rifle units, emphasizing aspects of the night attack most difficult to execute well.

PAGE 46

Merimskiy's book (43-45, 46), Krivosheyev ("Polevoy vyuchke - vysokoye kachestvo," 1985) and Groshev ("Nepreryvnyye nochnyye zanyatiya," 1988) discuss defense and MRB exercises (1984: 75-81, 90-97, 196-216; 1987: 291-314). Examples 4 and 8 refer to use of air defense weapons.

MOOD

Figure 1. Night attack by a Soviet battalion. Black and white pencil sketch. Nimble Books using Stable Diffusion.

Figure

2.

Figure 3. Exhausted Soviet soldiers after their battalion carried out a night attack. Black and white pencil sketch. Bill Mauldin style. Nimble Books staff art using DALL-E 2. The OpenAI product did much better at capturing the style of the famous WW2 American cartoonist than Stable Diffusion or Midjourney.

AD-A216 424

NIGHT ATTACK

BY

A SOVIET BATTALION

DTIC
S ELECTE
JAN10 1990
B D

SOVIET
ARMY
STUDIES
OFFICE

**Fort Leavenworth,
Kansas**

8905713

90 01 10 019

REPORT DOCUMENTATION PAGE

Form Approved
OMB No. 0704-0188

1a. REPORT SECURITY CLASSIFICATION	1b. RESTRICTIVE MARKINGS
Unclassified	
2a. SECURITY CLASSIFICATION AUTHORITY	3. DISTRIBUTION/AVAILABILITY OF REPORT
2b. DECLASSIFICATION/DOWNGRADING SCHEDULE	Unclassified/Unlimited

4. PERFORMING ORGANIZATION REPORT NUMBER(S)	5. MONITORING ORGANIZATION REPORT NUMBER(S)

6a. NAME OF PERFORMING ORGANIZATION	6b. OFFICE SYMBOL (If applicable)	7a. NAME OF MONITORING ORGANIZATION
Soviet Army Studies Office	ATZL: SAS	—

6c. ADDRESS (City, State, and ZIP Code)	7b. ADDRESS (City, State, and ZIP Code)
HQ CAC ATZL: SAS FT. Leavenworth, KS 66027-5015	

8a. NAME OF FUNDING/SPONSORING ORGANIZATION	8b. OFFICE SYMBOL (If applicable)	9. PROCUREMENT INSTRUMENT IDENTIFICATION NUMBER
Combined Arms Center	CAC	

8c. ADDRESS (City, State, and ZIP Code)	10. SOURCE OF FUNDING NUMBERS			
CAC Ft. Leavenworth, KS 66027	PROGRAM ELEMENT NO.	PROJECT NO.	TASK NO.	WORK UNIT ACCESSION NO.

11. TITLE (Include Security Classification)
NIGHT ATTACK BY A SOVIET BATTALION

12. PERSONAL AUTHOR(S)
GEBHARDT, JAMES F.

13a. TYPE OF REPORT	13b. TIME COVERED	14. DATE OF REPORT (Year, Month, Day)	15. PAGE COUNT
Final	FROM ___ TO ___	1989 JUNE	45

16. SUPPLEMENTARY NOTATION

17. COSATI CODES			18. SUBJECT TERMS (Continue on reverse if necessary and identify by block number)
FIELD	GROUP	SUB-GROUP	NIGHT COMBAT; NIGHT ATTACK; SOVIET NIGHT TACTICS (incl)

19. ABSTRACT

Soviet ground force tactical units conduct night attacks in accordance with a theoretical model which has changed little over the past decade. Its salient characteristics are pre-battle reconnaissance, attack from the march, dismounted assault, illumination, patrolling, commitment of a second echelon, and penetration of the defending brigade reserve positions by dawn. Demonstrated Soviet tactical unit deficiencies in executing the night attack include land navigation and terrain orientation, driving, and use of night vision devices. Theoretical vulnerabilities which may be exploited include over reliance on illumination, predictibility of employment of combat reconnaissance patrols, and physical exhaustion of Soviet troops. Keywords.

20. DISTRIBUTION/AVAILABILITY OF ABSTRACT	21. ABSTRACT SECURITY CLASSIFICATION
☒ UNCLASSIFIED/UNLIMITED ☐ SAME AS RPT. ☐ DTIC USERS	Unclassified

22a. NAME OF RESPONSIBLE INDIVIDUAL	22b. TELEPHONE (Include Area Code)	22c. OFFICE SYMBOL
Tim Sanz	913 684-4333	ATZL: SAS

DD Form 1473, JUN 86   Previous editions are obsolete.

NIGHT ATTACK
BY
A SOVIET BATTALION

by

Major James F. Gebhardt
Soviet Army Studies Office
U.S. Army Combined Arms Center
Fort Leavenworth, Kansas

June 1989

NIGHT ATTACK

BY

A SOVIET BATTALION

INTRODUCTION

Night combat [nochnoy boy] remains an important element of
Soviet tactics. Soviet experience in night offensive combat
dates to the Russian Civil War period. In November 1920, Red
divisions began a major offensive with a night attack to force a
broad water obstacle and seize a beachhead on the northern
approach to the Crimean peninsula.[1] This success led to the
defeat of Baron Wrangel's White forces and the eventual fall of
the entire peninsula. Examples of Soviet night offensive
combat in World War II abound, both in the European and Far
Eastern TVDs (theaters of military action).[2] In the postwar
years, based on their own wartime experiences, and on the study
of local wars and conflicts around the world, the Soviets have
continued to develop effective techniques for combat under
conditions of reduced visibility.[3] As a result, the Soviets have
a clearly-articulated theory for night offensive combat at the
subunit level (podrazdeleniye - battalion and lower).

This paper examines this theory, both as portrayed in
Soviet tactical studies, and through analysis of several night
attack exercises. It describes recognized Soviet training
weaknesses, and concludes with a description of potential
vulnerabilities, which might be exploited to defeat a Soviet
night attack.

SOVIET NIGHT ATTACK THEORY

Influence on combat actions

When analyzing Soviet night attack theory, one must first recognize Soviet military theorists' perceptions concerning factors inherent in offensive and defensive night combat.[4] There are numerous variables in the equation: the degree of darkness of the night; the amount of special equipment a unit has for night actions; the effectiveness of various types of equipment and weapons; and the level of training of units for night actions.

The Soviets acknowledge a wide range of technical equipment which can support the conduct of night actions, including electrical, pyrotechnic, night vision, radar, thermal, remote sensing, and signalling devices.[5] Electrical devices include searchlights, vehicle lights, flashlights, and so on. Pyrotechnic devices include all types of illumination ordnance, from ground trip flares to parachute flares launched from mortars, artillery, or aircraft. The Soviets describe night vision devices as both active and passive, and show a clear understanding of the advantages and limitations of each.[6] Radars include ground surveillance, artillery and mortar counterfire, and counter-radar equipment. Remote sensing devices include seismic intrusion detectors, remote sound sensors, infrared detection devices, and remote camera devices. Thermal sights and low-level-light television are described in terms of their use by

2

NATO in both ground and air systems.[7] Finally, the category
which the Soviets call "light signal devices" includes tracer
bullets, star clusters, ground flares, route and lane-marking
battery-powered lights, and lanterns.

The effect of these means are varied. Soviet theory posits
that soldiers not equipped with night vision sights require 1.3
to 1.5 times more ammunition for small arms engagements at night
than in daytime. The time required for motorized rifle units to
conduct a specific mounted maneuver is 30 per cent greater in
darkness. As Soviet sources emphasize, tank units can have great
shock effect on the enemy in night combat. Antitank and
artillery fires are less effective, due to problems of target
acquisition and fire adjustment. Ammunition and time expended in
night actions increase, and maneuverability of tank units
decreases. Artillery units need more time to occupy firing
positions, and expend more ammunition in fire missions.[8]

Air defense weapons which require visual target acquisition
and identification have greatly diminished effectiveness at
night, while those systems with electronic targeting capability
are not adversely affected. Aviation assets possess
effectiveness commensurate with the technological level of their
target acquisition/identification equipment. Engineer work using
night vision devices requires 25--30 per cent more time at night,
particularly since many engineer tasks must be accomplished with
heavy equipment, and under enemy observation. All support

3

activities, such as vehicle repair, medical evacuation, and resupply are also more difficult at night.

Soviet theorists recognize that conducting an attack at night accords several advantages. Units can engage the enemy with a less-pronounced superiority in forces than required for a daylight attack.⁹ They can more readily achieve surprise, and accomplish their missions with fewer losses than during the day. Attacking forces can reorganize and relocate under cover of darkness, thus retaining the initiative, and denying the enemy time to regroup or reinforce. Conversely, the Soviets identify a number of disadvantages. Even with technological advances in the development of observation and detection devices, it remains more difficult to maintain contact with the enemy at night. Darkness complicates terrain orientation and land navigation, and compounds routine and normal command and control problems. Mutual coordination [vzaimodeystviye] becomes more difficult to effect. The degraded performance of men and equipment, as previously noted, further complicates the attack. Finally, the Soviets acknowledge that night combat operations place greater physical and psychological stress on soldiers. They understand the importance of the physical and mental preparation of soldiers and units before they are committed to night combat.

General Principles

A night attack commences either from the march or from a position in contact with the enemy.¹⁰ These two forms differ

4

regarding the location of assembly areas and types of control measures employed. A unit attacking from the march usually passes through another unit in contact with the enemy, and thus gains the advantages of surprise and reduced vulnerability. A unit attacking from positions in contact generally is able to acquire more precise information on enemy disposition and strength, but is less likely to gain the element of surprise. Once the attacking unit has deployed and crossed the line of attack, the two forms are identical.

An attacking motorized rifle or tank battalion can perform a variety of roles at night, on either the main axis or in a supporting attack, in the first or second echelon, as a forward detachment, advance guard, or enveloping detachment. Its basic goal, according to the Soviet night combat specialist A. A. Rybian, is "the defeat of the opposing enemy and seizure of important positions or areas of terrain."[11] A Soviet reinforced tank or motorized rifle battalion attacking on the main axis at night normally operates on a narrower frontage than a similar-sized force not on the main axis.[12] While this force is expected to penetrate to the same depth as in the day, the mission is stated in a slightly different form. The immediate daytime mission of a reinforced battalion is normally to penetrate the defensive positions of the defending battalion and seize them.[13] The subsequent mission is to attack in concert with neighboring units and defeat the enemy brigade reserve. The last element of the mission is the direction of further advance.

If the defender has had time to develop fully his defensive fortifications, or is defending behind a river line, the immediate daytime mission of the Soviet reinforced battalion is to penetrate the defending company strongpoint and seize it. The subsequent mission is to penetrate and seize the defending battalion position with the aid of neighboring attacking units. At night, however, the battalion's immediate or subsequent mission is designated as the position to be reached by dawn. If because of the nature of the terrain, the strength of the defender, or the shortness of the night the battalion cannot achieve this depth, it must reach a specified line by dawn, and then continue the attack without pause in daylight.[14]

The Soviets consider many factors in establishing the combat formation [boyevoy poriadok] of a reinforced battalion for a night attack. Above all, it must have sufficient combat power in its first echelon to enable it to accomplish the immediate mission without reinforcement before dawn. To insure the greatest degree of tactical independence, tank/motorized rifle, artillery, and engineer reinforcments are attached to companies. During a breakthrough of prepared defenses, a battalion normally organizes with a first echelon of two companies, and a second echelon or reserve of one company. If the enemy position appears to be weak, or the attacking battalion is understrength, the second echelon or reserve can be as small as a platoon.

Batteries from the attached artillery battalion provide support to companies, with one platoon of artillery or mortars

6

set aside to fire illumination. At night more artillery, especially self-propelled, is used in the direct fire mode. Antitank guns and guided missiles deploy in sectors between first echelon companies and on flanks, as well as close behind the combat formations of the company on the main axis. Organic battalion air defense units disperse behind the first echelon companies, with attached air defense weapons farther to the rear, where they can protect the combat formation from enemy air attack, and also destroy enemy illumination support by shooting down parachute flares.[15]

The battalion second echelon, or reserve, moves closer to the first echelon at night, utilizing terrain for cover. This permits its rapid introduction into the battle. To avoid the problems associated with resubordination and reorganization at night, the second echelon normally receives its combat support attachments before dark.

Illumination in the Attack

Illumination in a night attack enhances the attacking force's activities, and at the same time complicates those of the defending enemy.[16] Illumination accomplishes several specific missions: exposing targets and objectives in the enemy's defenses, to include his reserve or counterattacking force; designating or marking routes of march or maneuver; marking the direction of the attack, and maintaining orientation; and marking zones of contamination, obstacles, bypasses, passage

7

lanes, and frontline traces. Illumination is also used to assist
in mutual recognition, designate targets, facilitate control, and
effect coordination with neighboring units. Illumination can
also blind enemy observation points, command and control
positions, and night vision devices used for observation or
weapons firing.

Soviet illumination support is always tightly organized and
controlled. Battalion commanders are encouraged to conserve
pyrotechnic illumination resources by judicious use of night
vision devices, so that they will have 15-20 per cent of their
illumination support in reserve for unexpected missions.[17]
Periodic illumination assists in terrain orientation,
observation, and the use of both direct and indirect fire
weapons. Continuous illumination, which requires significant
expenditures of resources, is applicable only to special
occasions, such as during the initial assault, while defeating a
counterattack, or during the introduction the second echelon.
the Soviets designate targets by direct illumination, by
illumination of a reference point with subsequent adjustment to
the target, or by signal rocket or tracer rounds.

Associated with illumination, the Soviets possess a wide
variety of nonluminous, luminous, and illuminated marking
devices, which they use for marking release points, routes,
lines of deployment, passage lanes, critical traffic points, and
so on. Combinations of colored lights differentiate between
units. Battalions employ prearranged infrared or visible light

signals to mark their positions for friendly aviation support.
Ordinarily, companies establish light posts (osvetitel'nyy post
rotoy)[18] to designate troop positions every 500 meters along a
battalion frontline trace. These markers are activated only on
order, and are extinguished as soon as the supporting flight
mission is completed. Several different methods assist in mutual
recognition of troops and equipment at night. Tank turrets and
BMP/BTR side panels carry white recognition markings, and
dismounted troops wear some type of color marking on their
sleeves or backs. Signal rockets, flashlights, infrared
emitters, and other devices provide challenge-response signals.

One of the most important Soviet uses of illumination is to
blind enemy troops or their night vision devices. They attempt
to do this with flares, searchlights, and incendiary-caused
fires.[19] Employment of blinding illumination is most critical
during movement, deployment, and assault on the initial defensive
positions. During their pre-battle reconnaissance, the Soviets
make every effort to locate the defender's command observation
posts and night vision devices. Conversely, enemy use of night
observation devices is a constant concern for the Soviet
commander, especially since newer-generation devices are mostly
passive, and thus are extremely hard to detect. Soviet
commanders and troops are taught the enemy organization and
equipment, enemy employment of night observation devices, and
recognition signs of their use.

Action against the enemy's illumination support is conducted in a detection phase and a destruction phase. Pre-battle reconnaissance seeks out enemy mortar and artillery firing positions for destruction during the Soviet artillery preparation. During the course of the battle, first priority is given to the destruction of any newly-discovered positions from which illumination is being fired against Soviet forces.

Preparation of the Night Attack

Soviet troop-leading procedures for the preparation of the night attack are essentially the same as for a daytime attack.[10] In fact, Soviet battalion commanders are trained to plan for the possibility that their daylight attacks may extend into the hours of darkness. Combat taking place at night, however, requires some additional planning elements. If the attack begins from a position in contact, the Soviet commander conducts his pre-battle planning and reconnaissance on the ground. If the attack is from the march, the commander normally makes a decision based upon a map study and, if possible, goes forward to look at the terrain, a process called rekognostsirovka, or commander's reconnaissance.

Commanders give special consideration to such things as time of year and day, which determines the duration of darkness and intensity of moonlight; the enemy's capability in the areas of illumination support and use of night vision devices; the illumination plans of adjacent Soviet units and how they will affect the attack; and the ability of the unit through which the

10

attacking unit will pass to support the attack with illumination.
The Soviet commander pays particular attention to weather,
because wind, precipitation, and humidity all affect his
illumination plan.

Organizing coordination for a night attack is both more
difficult and more important than for a daylight attack.[21] Night
attacks require additional control measures and recognition
signals, an illumination plan, a plan for the use of night vision
devices, a plan for the designation and use of illumination
reference points, measures to take in dealing with enemy
illumination and night vision devices, and a plan for transition
from night actions to day actions. The plans for the dispatch of
a post-breakthrough combat reconnaissance patrol [boyevoy
razvedyvatel'nyy dozor/BRD], the handling of the enemy's
counterattack force, and the employment of the second echelon or
reserve are not unique to night actions, but require careful
preparation as well.

CONDUCT OF THE NIGHT ATTACK (THEORY)

Soviet units conduct night attacks either from the march
or from positions in close contact with the enemy. These two
modes differ only in the initial stages. The attack from the
march normally begins with the movement of artillery units,
designated tanks, and other direct fire systems from the assembly
area into firing positions.[22] The artillery preparation, if
used, will often be shorter than during the day.[23] At the

11

appointed time, the battalion lead element crosses the start point and moves at predetermined speed to the lines of deployment into company and then platoon columns. Just before reaching the line of attack, platoons deploy on line into their combat assault formation. The degree of use of night vision devices depends on a number of variables, including level of ambient light, difficulty of terrain, and probability of enemy detection.

While the artillery fires the preparation, and the battalion deploys into company and then platoon columns, designated tanks, artillery, and other direct fire weapons conduct aimed fires on preselected targets, and the engineers clear lanes in enemy barrier systems. The Soviet battalion commander moves from his place behind the lead unit and occupies his command observation post, where he is joined by the supporting artillery battalion commander.

If the attack is executed from a position of close contact with the enemy, motorized rifle troops, with the exception of selected gun crews, snipers, and observers, remain under cover until the attack begins. Commanders man their posts, monitor the results of the artillery preparation and control their own direct fires. When the tanks approach the line of attack, motorized rifle forces in contact with the enemy occupy their forward trench positions and open fire on enemy antitank guided missile crews. With the aid of signal devices or markers, they show the tanks where to cross the trenches. As the tanks cross over the trenchlines, motorized rifle troops emerge from their positions and attack.

Whether attacking from the march or from a position in contact, if the enemy's initial positions are significantly distant, under normal terrain conditions motorized rifle troops attack mounted until they are 300-400 meters from the enemy. If enemy fire is weak, BTRs and BMPs move close behind the tanks, for protection from enemy tank and antitank fires. This lessens losses, saves the strength of the troops for the dismounted assault, significantly increases the speed of the attack, and facilitates closer interaction of motorized rifle units with tanks.[14] In terrain which is difficult for wheeled BTRs, infantry troops initially ride on tanks. Thus, tank commanders are informed immediately of enemy antitank fires.

As the initial assault begins, Soviet artillery continues to fire on known and suspected enemy positions, including the enemy's artillery and mortar firing positions and command posts. If illumination is being used at this stage in the battle, it will be continuous, ammunition stocks permitting. As the attacking force moves into the enemy's zone, artillery begins to fire marking rounds to establish and maintain the direction of the attack. All direct fire weapons engage targets exposed either by illumination or by their own fires. Enemy fixed or mobile searchlights are priority targets. If they cannot be destroyed, the Soviet commander uses smoke to blind them. As noted earlier, specially designated gun crews attempt to shoot

13

down enemy parachute flares, and antiaircraft gun crews engage air-delivered parachute flares (utilizing short range missiles in exceptional cases).

If the attack occurs without illumination, battlefield fires provide additional ambient light for night sights and observation devices. Quite often incendiary rounds are used to start grass or brush fires. If the enemy illuminates the battlefield, the Soviet commander will also illuminate, and continue the attack under illumination.

Tanks and BMPs cross minefields in their attack formation by using mine rollers and plows, while BTRs and other equipment use passage lanes.[15] Dismounted infantry follow in the tracks of the vehicles. If they cannot find or see the tracks, they cross in lanes marked by the combat engineers. If there are insufficient mine rollers, passage lanes are negotiated first by tanks, then by BMPs, and finally by other combat vehicles. The period of the negotiation of the enemy's barrier system is the most critical moment in the attack. During this phase the Soviet commander does his utmost to destroy or suppress enemy direct fire and observation positions. His antitank resources are especially important in this task, and are placed in overwatch positions.

Tank survivability is enhanced at night by their maneuverability, as well as by the difficulty of enemy gunners in maintaining a good sight picture in poor light conditions. Stabilization systems significantly aid tank crews in maintaining gun tube orientation, while still allowing tanks to maneuver.[16]

14

As tank and motorized rifle units advance into a well-illuminated zone, company and battalion commanders shift illumination fires forward to the next phase line.

Close coordination between motorized rifle and tank units is vital. Motorized rifle commanders, if they detect or are engaged by weapons systems which they are unable to destroy with their own means, designate targets for the tanks with signal flares or tracer shells. The motorized rifle units also use the smoke-generating capability of tanks to cover their movement.[17] Conversely, the motorized rifle units seek to destroy ATGM crews and positions as well as enemy illumination posts.

When his forces have penetrated the enemy position to the depth where his own artillery may endanger them, the Soviet battalion commander orders the shifting of fires and, if necessary, illumination. At this point, illumination is no longer continuous, but remains available. Soviet units take care not to bypass pockets of enemy resistance, since these groups pose a danger to the second echelon. Small forces from the first echelon neutralize these pockets. This is a departure from day-fighting techniques, when the second echelon forces frequently receive the specific mission to neutralize bypassed pockets of resistance.

At this stage of the battle, when the enemy's forward platoon and company positions have been defeated, the Soviet commander dispatches one or more combat reconnaissance patrols [BRD]. A battalion normally deploys a platoon, and a motorized rifle company a squad. The primary task of these patrols is to establish the location, size, and axis of the enemy reserve or counterattack force.[38] Companies of the first echelon which have accomplished their immediate mission attack on to the battalion immediate mission or, based on the information garnered by the reconnaissance patrol, are given new missions.

With the immediate mission fulfilled, the Soviet battalion commander introduces his second echelon company to increase his combat power and develop the offensive into the depths. Soviet tactical doctrine posits that only 13-18 minutes will be required for this action (7-10 minutes for alert, designation of mission, and resubordination of support, and 6-8 minutes for movement and deployment).[39] The battalion commander personally directs the commitment of the second echelon company, and gives the mission to its commander orally at his command post or by radio. The second echelon company's attack is often supported by the fires of first echelon companies, which also continue to attack within the enemy positions. Whenever the second echelon is committed, a reserve is created by pulling back a company or platoon from the first echelon.

Having penetrated the enemy's forward battalion defensive positions, the Soviet battalion swiftly deploys on the specified axis in pre-battle formation [predboyevoy poryadok][30] to catch and destroy the retreating enemy before he can set up a cohesive defense at a subsequent position. If the enemy does succeed in reestablishing a position, the Soviet commander tries to fix it

16

with artillery fire, then attacks it from the flanks or rear.
The combat reconnaissance patrol (BRD) receives the mission to
find exploitable gaps between defensive positions or on the
enemy's flanks.

If the BRD has detected an enemy counterattack force, the
Soviet battalion commander considers whether to attack it in a
meeting engagement, or defeat it by some combination of direct
and indirect fires from a position on favorable terrain. It is
vital to prevent any coordinated action between the fixed enemy
force and his counterattacking force. If necessary, the Soviet
battalion commander moves a unit, perhaps his antitank reserve,
across the enemy route of advance, and orders his combat
engineers to lay a hasty minefield in conjunction with an ambush.

Just as in the day, if the enemy counterattack force is
superior to the attacking Soviet battalion, the main body quickly
establishes defenses on a favorable line in single echelon, with
a small reserve.[11] An obstacle plan is implemented, artillery
fire plans are coordinated, motorized rifle troops occupy
dismounted positions, and vehicles are emplaced in mutually
supporting firing positions. Antitank weapons deploy on the main
armor avenue of approach, and the automatic grenade launcher
platoon is sited on the avenue of greatest dismounted threat.

Concentrated indirect and direct fires under illumination
strike the counterattacking enemy first. Should any enemy tanks
penetrate the defense, they are destroyed by antitank fires, the
reserve, or artillery direct fire. The Soviet commander may send

out small teams of tanks and infantry to hit the counterattack force in the flanks or rear. When the counterattack is finally defeated, the Soviet battalion continues its attack, alone or in concert with adjacent units, and defeats the enemy. When the Soviet battalion attains its subsequent mission, it occupies positions and resupplies in preparation for its next mission. This mission (usually at dawn) may be to defeat another counterattack, or to continue the attack. In either case, the battalion is to be prepared to act without pause between night and day combat actions.[32]

CONDUCT OF THE NIGHT ATTACK (PRACTICE)

Having examined in some detail the Soviet theory for night offensive operations, it is now appropriate to look at how they implement that theory. Several exercises involving motorized rifle or tank battalions, which occurred over the fifteen-year period from 1974 to 1988, have been described in the open press. Specific characteristics or features were common to them all (See Table 1 and Figure 1).[33]

Six of the nine exercises involved a motorized rifle battalion base, the remainder employed a tank battalion. In all nine examples, the most common feature was the employment of a reinforced motorized rifle or tank battalion supported by an artillery battalion. In seven of nine cases, a platoon or squad of engineers was attached, and used in standard fashion for route/lane marking and/or obstacle clearing. When air defense

18

Figure 1

2d Motorized Rifle Battalion Attacks at Night

(map depicts attack in Example 7)

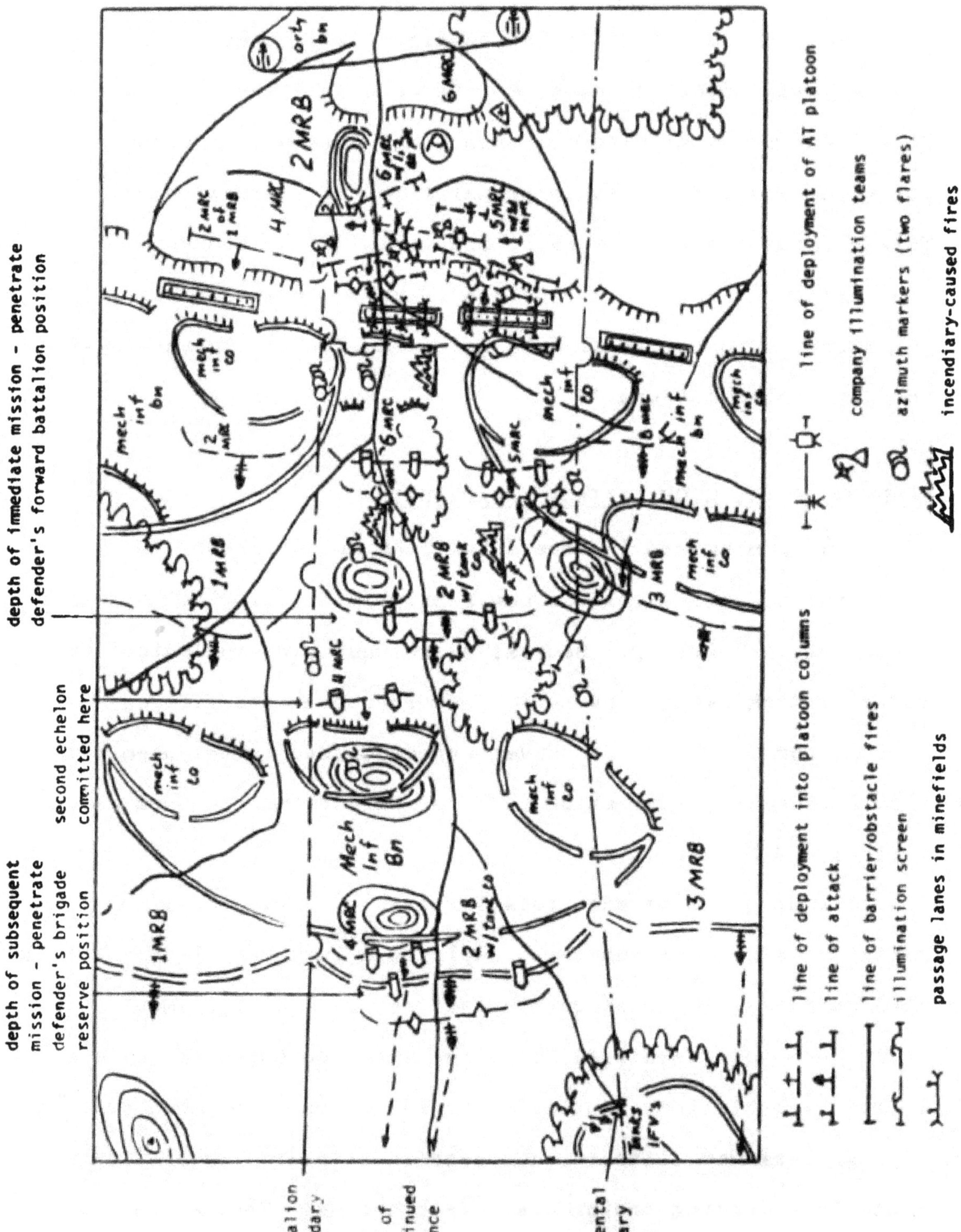

Table 1

Example	1	2	3	4	5	6	7	8	9
Attacker	MRB	MRB	MRB	TB	MRB	TB	MRB	TB	MRB
Defender	company	company	battalion	battalion	battalion	battalion	battalion	battalion	battalion
Missions: Immediate	Attack from march defeat co strongpoint	Attack from contact defeat co strongpoint	Attack from march defeat bn position	Attack from march defeat bn position	Attack from march defeat bn position	Attack from contact penetrate to bn rear	Attack from contact defeat bn position	Attack from march defeat bn position	Attack from contact defeat bn position
Subsequent	defeat bn reserve	defeat bn reserve	defeat enemy second posn	defeat bde reserve	seize a line		defeat bde reserve	defeat bde reserve	defeat bde reserve
Support Assets	Arty Bn Tk Co AT Plt En Plt NBC Sqd	Arty Bn Tk Co Aviation AT Plt	Arty Bn Tk Co En Plt AT asset AGS Plt	Arty Bn Mor Bat MRC En Sqd NBC Sqd ADA Plt	Arty Bn Tk Co AT Plt En Plt AGS Plt ADA Plt Helos	Arty Bn MRC ADA	Arty Bn Tk Co En Plt NBC Sqd AGS Plt ADA Plt	Arty Bn MRC En Sqd ADA Plt	Arty Bn Tk Co En Sqd ADA Helos
Art prep	20 min	10 min	yes	yes	12 min	yes	40 min	12 min	yes
Illumination	arty mortars LP's	avn mortars LP's	arty mortars	arty mortars LP's	NOD's, illum on contact	arty LP's	arty LP's	NOD's, illum on contact	arty mortars LP's
Dismounted Assault	yes	yes	unknown	yes	yes	unknown	yes	unknown	yes
Echelonment*	8/4	9/3	8/4	8/4	8/4	8/4	9/3	8/4	9/3
Smoke		arty TDA	mortars	fires	arty TDA fires	arty fires	fires	TDA	arty fires
Combat Recon Patrol (BRD)	3, 1 ea MRC	1 from 1st ech	2 from 1st ech	1 from reserve	1 from 1st ech		1 from unspec	1 from 1st ech	1 from unspec
Counter-attack	yes	yes	yes	yes	yes	yes	no	no	yes
2nd ech/res employment	on main axis	on main axis	to defeat X-atk	on main axis	at dawn to cont attack	on main axis	on main axis	on main axis	on main axis
	1st ech defeats X-atk	1st ech defeats X-atk		1st ech defeats X-atk	1st ech defeats X-atk				1st ech defeats X-atk

Legend:
AT	antitank	MRB	motorized rifle battalion
ADA	air defense artillery	MRC	motorized rifle company
AGS	automatic grenade launcher	TB	tank battalion
LP	light post	TDA	termodymovaya apparatura [smoke-generating device] (tank-mounted)
NOD	night observation device		

* Expressed in maneuver platoons per echelon. One echelon plus combined arms reserve in 1 and 4, two echelons in remainder.

elements were attached, they were always given the mission to engage enemy illumination as well as aircraft. When present, the automatic grenade launcher platoon was attached to the first echelon forces or placed immediately behind them under the control of the battalion commander. Although an NBC reconnaissance element was present in some cases, there was no particular attention given to it, indicating that it performed its standard mission.

In a majority of cases the immediate mission was to attack from the march to penetrate an enemy battalion-sized defensive position. In a third of the cases, the map depicted the Soviet force attacking with boundaries narrower than the defending battalion, thus only partially engaging the two forward defending companies. In another third of the cases, the Soviet force attacked with boundaries which coincided with the defending battalion's boundaries. The subsequent mission, either stated or indicated by maps, was generally to penetrate the enemy brigade reserve in conjunction with adjacent attacking units. Two points become obvious from these examples. First, there was never a statement of the mission in terms of width or depth of penetration in kilometers, and the orientation was never on a terrain feature, but always on an enemy force. Second, rarely did a single Soviet battalion conduct a night attack in isolation. Generally it was depicted, albeit notionally, as part of a regimental-sized or larger night offensive.

Every attack was preceded by an artillery preparation
lasting at least ten minutes. Initial fires were always
directed against known and suspected targets on the forward edge
of the enemy defenses. Therefore it is reasonable to conclude
that the Soviets rarely conduct night attacks by a battalion-
sized element without artillery preparatory fires and adequate
reconnaissance. The same can be said of illumination. In two
examples, the Soviets went into battle using night vision
devices, with illumination on call. In one case (Example 5), the
enemy's early detection of the attack forced the Soviet commander
to revert to illumination for the initial assault. In the other
case (Example 8), the tanks and BMPs used night sights, but still
relied on illumination reference points to maintain direction.
Illumination was delivered by airplanes in Example 2 and
helicopters in Examples 5. In all cases save one (Example 3),
motorized rifle troops established light posts. In most cases,
therefore, Soviet night attacks are initiated under illumination,
and in all cases illumination is available and used if needed.

In six of nine attacks, the Soviet force conducted a
dismounted assault on the enemy positions. In the remaining
three cases, the text referred to "line of attack", but did not
indicate that troops dismounted. One can conclude from this that
in most, if not all, night attacks, the Soviets plan for and use
dismounted motorized rifle troops in their initial assault on
enemy positions. Common to two-thirds of the attacks in this
analysis was Soviet use of smoke, either to mask their own

22

movements, or to blind the enemy. Smoke was generated from three principal sources: artillery/mortar shells, natural fires started by incendiary shells, and on-board tank smoke generators.

Regarding force echelonment, analysis indicates that in all cases the first echelon contained at least two-thirds of the combat power. This clearly supports the principle of giving the first echelon enough strength to accomplish its mission without reorganization. Commitment of the second echelon or reserve was always planned and carried out, either on the main axis to maintain the tempo of advance, or to defeat the enemy counterattack. Eight of nine Soviet battalion commanders sent out at least one combat reconnaissance patrol (BRD). This patrol was normally a platoon from the first echelon, which had the mission to find the enemy counterattack force. One commander was taken to task for sending out just one BRD (Example 7). In seven of nine cases, the enemy conducted a counterattack, which was defeated by a combination of artillery fire, direct fire, and maneuver. The Soviet commander always considered the likelihood of a counterattack in his plan, and was never surprised when it occurred.

Based on the analysis of these nine examples, a composite Soviet night attack can be constructed which in every way reflects the theoretical model. The attacking force is a reinforced battalion of combined arms composition organized into two echelons, with a heavy first echelon and a second echelon or reserve. The attack is preceded by an artillery preparation and

conducted under full illumination. The initial assault on the
enemy position contains a dismounted element. A combat
reconnaissance patrol deploys to locate the counterattack force,
which in turn is defeated by a combination of fire and maneuver.
The second echelon or reserve is normally committed, either on
the main axis or against the counterattack. The Soviet
reinforced battalion is generally prepared to continue the attack
at dawn.

VULNERABILITIES OF A SOVIET NIGHT ATTACK

Analysis of the theory and practice of the Soviet reinforced
battalion night attack identifies several training weaknesses and
theoretical vulnerabilities. The Soviets openly discuss training
weaknesses in their tactical-level military periodical journals.
Theoretical vulnerabilities, on the other hand, are rarely
discussed in the Soviet open press.[34]

Training deficiencies

Training deficiencies in night operations are a recurring
theme in the Soviet military press. One critic, a major general,
observed in a 1979 article:

> Night exercises are conducted irregularly, the
> proven principle of training - from the simple to the
> complex - is not observed. There is not a strictly
> adhered to system in the training of troops for night
> operations. The troops have been poorly conditioned
> in the skills of battlefield orientation and
> observation, in determining target range and
> designation, in the skillful utilization of night and
> day sights. Driver-mechanics and vehicle commanders
> have poorly mastered the driving of BMPs, BTRs, and

24

tanks using night vision devices, and under conditions
of blacked-out lights.[36]

The author then criticized commanders who did not conduct active

combat operations in night exercises, but instead limited their

activities to a march or reinforcement of an already-seized

position. Some commanders rarely used artillery and mortar units

for illumination support of night exercises and training. When

illumination was employed, it was often delivered inaccurately,

and served only to unmask friendly troops and interfere with

their use of night vision devices. The author complained that

tank gunnery training was poorly organized, sometimes extending

into the daylight hours, and thus denying units the practice they

needed in night firing. He further criticized the "timid" use of

navigational devices.

He indicated that officers and sergeants were insufficiently

trained in the tactical and technical characteristics of

illumination, signal, and night vision devices, and as a result

experienced serious difficulties conducting and adjusting fire.

Norms pertaining to tactical training, weapons firing, technical

training, defense against weapons of mass destruction, engineer

training, military topography, field medical training, and other

subjects of training at night exercises were poorly developed in

a number of cases. Finally, according to the author, there were

some problems in the training facilities themselves.

> Not everywhere yet does the training material-
> technical base support the qualitative and effective
> training of troops for the conduct of night operations.
> In some places the demands for properly prepared and
> equipped target areas on range complexes, and moving

25

target ranges, especially for the conduct of live fire tactical exercises, have not been completely fulfilled.[36]

Another article in the same journal addressed inadequacies in tank gunnery training.[37] The author observed that soldiers displayed poor observation and target detection skills, an inability to "sense" rounds [universal tanker jargon for observing the precise strike of projectiles in the target area, necessary for adjustment of subsequent shots -JG], and failure to make subsequent adjustments. He blamed inadequate leadership for these observed training weaknesses. Specifically, officers and sergeants did not organize and conduct the exercise properly, they did not devote sufficient attention to the principles of night firing and, worst of all, they did not observe and critique the gunners on an individual basis.

Tank gunnery training was also the subject of a 1980 article, in which a major general criticized the sporadic, occasional occurrences of night exercises, the failure to observe methodological sequences in training, the low organizational level of night gunnery training in some units, and the inadequate attention given to the achievement of norms under night conditions.[38] He also criticized tank and BMP night-driving training, in that headlights and searchlights were being used instead of night vision devices. He criticized vehicle commanders and drivers for their inadequate mastery of the BMP navigational apparatus. On a larger scale, exercises were poorly planned, leaving little daylight to the commander for terrain reconnaissance. Some units, due to the fear of the commanders to

26

conduct any kind of maneuver in darkness, either acted passively, or conducted only straight-line movement. These criticisms were quite similar to those made eighteen months earlier by another general officer.

A 1981 article again noted driving and navigational problems in an exercise.[39] During a night attack by a motorized rifle battalion, two platoons of the company on the main axis strayed off course due to a loss of orientation.

Night training was the subject of the lead article in the August 198? edition of Voyennyy vestnik [Military herald]. The author, Commander of the Carpathian Military District, addressed several training deficiencies in his units.

> In these units night training is organized
> simplistically, without a strict system. Tactical
> drill excercises planned for darkness frequently
> carry on into the day. Navigational apparatuses are
> not used. Night training, group exercises, and tactical
> flights with sergeants, warrant officers, and officers are
> conducted irregularly. As a result, many of them do not
> have simple skills in organizing combat under these
> conditions, and are not able to accomplish the reliable
> destruction of the enemy by fire, conduct coordination,
> exercise command and control of subunits and fires, and
> organize illumination support.[40]

The author went on to say that in night exercises, units were often totally inactive or accomplished only simple movements, and once again surfaced general inadequacies in night driving and firing.

A January 1983 article provided additional evidence that night orientation/land navigation is a serious and widespread problem in Soviet tank and motorized rifle units.[41] The article addressed the training of officer cadets at the Leningrad

Advanced Combined Arms Command School in the proper conduct of a
night attack by a BMP platoon from the march. In the practical
application portion of this lesson, the squad on the main axis of
the platoon failed to observe strictly the illumination reference
point, and thus the platoon did not maintain its axis of attack.
In addition, the cadets demonstrated weak knowledge of SOPs, had
difficulty working with maps, were weak in terrain orientation,
and had not mastered the use of BMP navigational apparatus. The
significance of these officer cadets' training weaknesses is
multiplied by the crucial role junior officers play in individual
and small unit training in the Soviet Army, which lacks a
professional noncommissioned officer corps.

Several months later, another lead article criticized
training proficiency in night operations in two widely separated
military districts.⁴³ In a night motorized rifle unit exercise
in the Baltic Military District, command and control was "shaky",
illumination support was "primitively" organized, company light
posts were not established, and many troops had not learned how
to utilize their equipment and weapons, in particular how to
employ illumination, to use light-signalling and night vision
devices, or how to move on an azimuth during periods of darkness.
In the Central Asian Military District, battalion tactical
exercises did not always include dynamic night combat operations.
The article singled out one particular company-level exercise,
where the exercise director assigned a vague mission, and

provided neither illumination nor communications equipment. As a result, he lost control of the training unit.

More recently, a 1987 article describing a company-level night attack exercise in Group of Soviet Forces Germany listed as common training deficiencies the following: maintaining orientation and direction; using night vision devices and means of illumination; controlling weapons firing; and maintaining coordination among units.[43]

Thus, Soviet military professionals admit to a number of deficiencies in the training of their leaders and soldiers in night operations.[44] Principal among them are orientation and navigational difficulties, insufficient individual training of soldiers in skills pertaining to employment of weapons and equipment, planning and execution of illumination support, and the ability or even willingness of commanders to maneuver their units aggressively during conditions of darkness. Having admitted these deficiencies, the Soviets appear to be making continuous efforts to correct them. In 1984, Colonel General V. Merimskiy, at that time Deputy Commander-in-Chief of Ground Forces for Combat Training, stated that "Instruction of subunits in night operations is one of the most important missions of combat training."[45] The two examples he presented as models of night training were a squad in the defense, and a motorized rifle battalion in the defense which went over to the attack.[46]

A July 1985 article provided another indication of serious interest in night training. Writing about the field training program of Group of Soviet Forces, Germany, Colonel General

Krivosheyev, Chief of Staff, stated that not less than 30 per

cent of field training time was dedicated to night training.[47]

As an example he referred to a one-night training exercise for

tank units, in which they were alerted, moved to a suitable

training area, and put through a controlled live-fire exercise to

verify their combat proficiency in night fire and maneuver.

Another article in 1988 detailed the frequency of night training

for motorized rifle units in the Group of Soviet Forces, Germany.

Squads and crews train once a month for three-four nights;

platoons once quarterly for four-five nights; companies and

batteries once in each training period for five-seven nights; and

battalions once annually for seven-ten consecutive nights.[48] But

because of the constant turnover of enlisted personnel in Soviet

units, and the absence of an experienced noncommissioned officer

corps, these problems, which appear to be typical both over time

and geography, will never completely go away.

Theoretical vulnerabilities

A theoretical vulnerability is an aspect of the night

attack which, in the author's opinion, even if executed according

to established norms, might render the Soviet tactical operation

vulnerable to detection, disruption, or defeat. Vulnerabilities

range throughout the preparation for and conduct of a night

attack. The first period of vulnerability occurs during the

preparatory phase, normally still during daylight, when the

30

Soviet commander is conducting reconnaissance with his subordinate commanders. Their activities are directed primarily at pinpointing the disposition of enemy forward positions, command posts, and firing positions of direct and indirect fire weapons. To reduce this vulnerability, the Soviets, whenever possible, use the assets of the unit in contact.

Soviet reconnaissance efforts can be countered in three ways. Opposing maneuver forces must be alert to the presence of Soviet reconnaissance parties. When detected, they can be engaged by indirect or direct fire. Reconnaissance parties should be eradicated from temporary or alternate direct fire positions, to prevent disclosure of primary positions. A second counter to Soviet reconnaissance is to deny them the opportunity to collect information. Defending units should observe good principles of operational security (OPSEC), especially those which pertain to camouflage and concealment. Since the Soviets will always be able to observe a certain amount of activity, the third counter is to resort to additional measures to complicate Soviet reconnaissance. A defending unit should provide false information, by constructing dummy barriers and firing positions, operating false radio nets, and moving combat vehicles and units in such a manner as to portray its disposition falsely. Such activities may be beyond the resource capability of a company/team commander and, therefore, require the support of higher command levels. Every defensive plan should contain within it an OPSEC and a deception plan, coordinated within the context of a higher deception plan, which will help to defeat Soviet reconnaissance activities.

The next period of potential vulnerability occurs when the Soviet attacking force begins the prepositioning of indirect and direct fire weapons systems. In an attack against prepared positions, the Soviets must preposition regimental and divisional artillery units used to fire large-scale preparations. This movement must be detected in order to provide early warning to the defending unit and the possibility of implementing countermeasures. A NATO defending battalion task force commander probably has access only to assigned or attached ground surveillance radar, and therefore has to rely on information acquired by the technical resources of higher headquarters. For immediate intelligence, the defending battalion commander relies on reports from his own frontline soldiers. Their ability to recognize Soviet tanks, BMPs, and self-propelled artillery by sight may be good, but how many can identify Soviet vehicles by sound? At night, when limited visibility will greatly impede the visual observation of Soviet movement, NATO soldiers have to rely on other senses which may not have been trained.

The decision to disrupt a Soviet night attack during the positioning phase must be taken only after consideration of Soviet intentions versus the risk of disclosure of one's own defensive positions. The possibility always exists that the Soviet commander is maneuvering for the sole purpose of drawing fire, so as to acquire targeting data. The defending commander

32

must weigh this factor against other indicators, including his own preparedness to withstand an attack. Because Soviet night attacks tend to be conducted from the march, there will be a high density of combat vehicles and units just at that point where the attacker passes through the forward unit. If the defender can detect this moment in both time and space, he can inflict serious damage on concentrated enemy forces with any combination of surviving appropriate weapons.

If the Soviet commander for any reason does not reposition any units or fire support assets, or the defending commander does not take note of Soviet repositioning, the first indication of a night attack will be the artillery preparation, during which Soviet engineer troops force passage through the defender's barrier systems. Defending units must therefore be physically and mentally prepared to engage these engineer troops, even as the artillery rounds fall around them. They must be well entrenched, with overhead cover, and be conditioned to keep the barrier system in their sights. The artillery preparation itself renders Soviet artillery and mortar firing positions vulnerable to counterbattery fires. Given the Soviet dependence on indirect fires for illumination and combat support, effective counterbattery fires could disrupt the attack in its initial stages.

Because NATO combat units probably enjoy a technical superiority in night vision devices, and the Soviets prefer to attack under full illumination, defending units should not strive

to "out-illuminate" a Soviet night attack. Their artillery
effort would be much better spent in firing high explosive
rounds at the attacking enemy and his sources of illumination
support. But there is something to be gained in throwing some
illumination back at the attacking force. For the defending
troops, even though they are equipped with a plethora of
relatively effective night vision devices, illuminating the
enemy provides a great psychological boost. Since not every
weapon, especially at the individual level, is night-sight
equipped, illuminating the enemy will also improve the
effectiveness of the defenders' small arms fires. A second
advantage of this technique is that it causes Soviet air defense
systems to engage parachute flares with both guns and missiles.
Soviet doctrinal and training publications stress this mission
for air defense assets.⁴⁹ Every bullet, shell, or missile fired
at a flare is one less fired at a ground or aviation target. It
also discloses the location of Soviet antiaircraft weapons, which
can then be targeted. This tactic should be considered prior to
the employment of close air support by the defending unit.

One of the major training deficiencies of Soviet units is
that of night land navigation and terrain orientation. A
defender can exploit this vulnerability by imitating Soviet
navigational aids. Soviet units maintain direction (azimuth)
through periodically fired artillery marking rounds on the main
axis. If the defending unit duplicates the type of round being
fired, be it an illumination round, a smoke air burst, or a star

34

cluster, the Soviet force possibly can be misoriented. Imitative deception through pyrotechnics may also be used to confuse the Soviets as to the precise location of their own interunit boundaries.

If the attack succeeds in penetrating the initial defensive positions, the Soviet commander sends out one or more platoon-sized combat reconnaissance patrols (BRD). The defending commander must plan to detect and destroy these patrols, and in so doing deny the Soviet commander knowledge of the disposition of reserves, counterattack forces, or subsequent positions. How best to defeat the BRD? Leaders down to platoon level must know of its existence and mission, and anticipate its use. Company/team commanders should have a contingency plan to neutralize the BRD when it is discovered in their zone. Battalion task force commanders must actively seek out BRDs, and plan for their destruction, either by fire or maneuver forces, keeping in mind the Soviet tendency to utilize minefields and ambushes. If the Soviet commander is denied information on the disposition of the defender in depth, he has to slow the tempo of his attack. He is then more vulnerable to interdicting artillery or aviation strikes, and the defender has more time to consolidate his defense.

According to exercise experience, the Soviet commander normally commits his second echelon or reserve to gain his final objective line or continue the attack at daylight. The defending commander should, therefore, anticipate its use and plan accordingly. He should attempt to discern as early as possible its size, composition, and route and location of deployment, keeping in mind that the Soviet commander introduces it after the initial company positions have been penetrated in a prepared defense, or after the initial battalion position has been penetrated in a hasty or partially prepared defense. Based on this information and forces available to him, the defending commander can plan a response to defeat this fresh Soviet attacking force.

Finally, the defending unit must do everything possible to force the Soviet attacker to dismount early and stay dismounted. Obstacles must be placed in front of, between, and behind initial defensive positions. Keeping Soviet troops out of their vehicles will exacerbate their already acute command and control problems. It will greatly diminish their rate of advance, physically exhaust them, and make them more vulnerable to artillery and small arms fires.

CONCLUSIONS

A night attack by a Soviet motorized rifle or tank battalion is conducted according to a theoretical model which has changed little over the past decade. Its salient characteristics are active pre-battle reconnaissance, attack from the march, dismounted assault under continuous illumination, deployment of combat reconnaissance patrols into the defender's depth, commitment of a second echelon, and penetration of the defending brigade reserve positions by dawn.

Soviet troops have displayed numerous night training deficiencies, most notably in land navigation and terrain orientation, driving, and use of night vision devices. Given the conscript nature of their force, and current two-year term of service, there are limited opportunities to undertake continuous training to overcome these deficiencies. Therefore, in the absence of substantial change to terms of service, force composition, or pre-induction/active duty/reservist training programs, the general proficiency of the Soviet Army in night operations is not likely to improve. Added to these training weaknesses are theoretical vulnerabilities which, if properly exploited, will aid in the defeat of the Soviet night attack. These vulnerabilities include susceptibility to deception during reconnaissance, over-reliance on illumination, susceptibility to imitative deception by pyrotechnics, predictibility of the employment of combat reconnaissance patrols and second echelons, and physical exhaustion of Soviet troops if they can be kept dismounted.

Clearly, surviving and defeating a Soviet night attack will depend on a study of Soviet tactical theory, an understanding of Soviet vulnerabilities, and the development of plans and methods well in advance of the requirement to use them in a future conflict.

1. N. V. Ogarkov, ed., Sovetskaya voyennaya entsiklopediya [Soviet military encyclopedia], (Moscow: Voyenizdat, 1978), 6:286--87, sv (Perekopskaya-Chongarskaya operatsiya [Perekop-Chongar operation]).

2. A particularly useful survey of Soviet night combat in World War II is contained in the article by B. Panov, Osobennosti vedeniya boyevykh deystviy noch'yu po opytu voyny [Special features of the conduct of night combat actions according to war experience], Voyenno-istoricheskiy zhurnal [Military historical journal, hereafter cited as VIZh], No. 10 (1980), 10--17. For a more recent treatment in English, see Claude R. Sasso, "Soviet Night Operations in World War II" Leavenworth Paper No. 6, (Ft. Leavenworth, Kansas: Combat Studies Institute, 1982).

3. See P. Tsygankov, Razvitiye taktiki nastupatel'nogo boya noch'yu v poslevoyennyye gody [The development of the tactics of offensive night combat in the postwar years], VIZh, No. 10 (1978), 53--61.

4. The principal source for this entire section on Soviet tactical theory is A. A. Rybian, Podrazdeleniya v nochnom boyu [Subunits in night combat], (Moscow: Voyenizdat, 1984), chapters 1 and 2, hereafter cited as Rybian, Podrazdeleniya. Material found in Rybian is summarized in V. G. Reznichenko, ed., Taktika [Tactics], (Moscow: Voyenizdat, 1984), 136--43; the same summary is reprinted with minor changes in the 1987 edition of Taktika, 260--69. D. A. Dragunskiy, ed., Motostrelkovyy (tankovyy) batal'on v boyu [Motorized rifle (tank) battalion in combat], (Moscow: Voyenizdat, 1986), 81--85, contains a short section on the night attack which has no substantial changes. A. S. Noskov, ed., Motostrelkovaya (tankovaya) rota v boyu [Motorized rifle (tank) company in combat], (Moscow: Voyenizdat, 1988), 82--83, likewise contains a short section on night attack.

5. Rybian, Podrazdeleniya, 21, section entitled "Means Which Support the Conduct of Combat Actions at Night". Intentionally or not, Rybian omitted the mention of lasers, which have great utility in modern night combat. Lasers can be used to designate targets, determine range, and disable or destroy night vision devices, as well as the vision of human observers.

6. 'Active' night vision devices are those which emit some type of signature, such as infrared searchlights, weapon sights, and drivers' periscopes. 'Passive' night vision devices are those which do not give off a signature, but instead, for example, amplify ambient light (starlight scope) or sense heat differential (thermal sight).

7. Rybian, Podrazdeleniya, 31--33.

8. For tanks, the Soviets use the figure of 30--50 per cent more time and ammunition, and 30 per cent degradation in maneuverability at night. For artillery, the corresponding figures are 25--30 per cent and 20--30 per cent. Ibid., 12--13. Just from the ammunition requirements alone, one can see that extended or continuous night combat places a significant additional burden not only on the individual soldier, but also on the logistic system.

9. On a much larger scale, this was the case with the 8th Guards Army in October 1943, attacking strong German defenses in front of Zaprozh'ye. Facing powerful 'Tiger' tanks and 'Ferdinand' assault guns, and lacking adequate supplies of artillery ammunition, the army commander first ordered the night infiltration of assault groups to destroy German armor, then a night offensive by his entire army. Attacking at night allowed the army commander to use a short (ten-minute) artillery preparation, and at the same time denied the Germans the ability to employ their superior armor firepower. See V. I. Chuykov, V boyakh za Ukrainu [In battles for the Ukraine], (Kiev: Politizdat Ukraine), 96--97.

10. In Russian, s khodu. This does not at all imply a meeting engagement, but rather that the Soviet unit is coming forward from an assembly area or previously-gained objective, with control measures enroute to guide its commitment into combat against an enemy force which has established a hasty or prepared defensive position.

11. Rybian, Podrazdeleniya, 38.

12. A reinforced tank or motorized rifle battalion will as a minimum include an artillery battalion, a motorized rifle company for a tank battalion, a tank company for a motorized rifle battalion, and other combat and combat support attachments as required (see Table 1 for examples from exercise experience).

13. For a more complete discussion of Soviet mission depths, see Lester W. Grau, "Changing Soviet Objective Depths: A Reflection of Changing Combat Circumstances," Soviet Army Studies Office, Fort Leavenworth, Kansas, March 1989.

14. Rybian, Podrazdeleniya, 42.

15. Ibid., 45. Although the Russian here is somewhat ambigious [unichtozheniye ego sredstv svetovogo obespecheniya], the point is made unequivocably later in the text: V sluchaye primeneniya protivnikom dlya osveshcheniya svetyashchikh aviatsionnykh bomb oni porazhayutsya ognem zenitnoy artillerii i krupnokalibernykh pulemetov, a v isklyuchetel'nykh sluchayakh i zenitnymi raketami

maloy dal'nosti deystviya [In the case of enemy utilization of aircraft-dropped flares for illumination, they can be destroyed by antiaircraft artillery and large-caliber machineguns, and in exceptional cases also by short-range antiaircraft missiles.] (87) Exercise analysis confirms this use of air defense weapons.

16. Ibid., 47. For a detailed illumination plan for a motorized rifle battalion's night attack, see Yu. Struchkov, and A. Yatsenko, "Svetovoye obespecheniye boya" [Illumination support of battle], Voyennyy Vestnik [Military Herald, hereafter cited as VV], No. 3 (1988), 26--28.

17. Rybian, Podrazdeleniya, 50.

18. A good English equivalent would be "illumination team". Using hand-fired rockets of various sizes, a company light post can illuminate an area of immediate concern to the company commander. Normally one or two will be specified for each company attacking in the first echelon.

19. Despite their utility in this role, Soviet sources do not discuss the use of laser-emitting devices in this context. Incendiary-caused fires are a multi-purpose weapon. The light illuminates the surrounding terrain, favoring the attacker. The flames themselves may ignite stored fuel, ammunition, camouflage, fortifications, and combat vehicles. The smoke produced can mask enemy observers, and disable or distract individual enemy soldiers. The heat generated may render less effective the defenders' thermal sights. The Soviets, and the Russians before them, have a long history in the use of fire as a weapon in war. Incendiary-caused fire is noted in the historical example which accompanied the night attack described in V. Godun, Batal'on nastupaet noch'yu [The battalion attacks at night], VV, No. 11 (1982), 27.

20. A. S. Noskov, ed., Motostrelkovaya (tankovaya) rota, 82.

21. Rybian, Podrazdeleniya, 79.

22. Ibid., 83.

23. Soviet sources do not explain why the artillery preparation is generally shorter at night, given that more time and ammunition are required for effective fire. Perhaps the elements of shock and surprise, which favor a brief preparation, are deemed as important as target effect, which would require a longer preparation.

24. Rybian, Podrazdeleniya, 86.

25.25. A BMP-2 with mine plow is pictured in Soldat und Technik (December 1988), 736.

26.26.   Beginning with the T-55, all Soviet tanks have
stabilization systems.

27.   All Soviet tanks beginning with the T-55 have a
<u>termodymovaya apparatura</u> [smoke-generating apparatus]

28.   The BRD is discussed more fully in F. I. Gredasov,
<u>Podrazdeleniya v razvedke</u> [Subunits in reconnaissance], (Moscow:
Voenizdat, 1988), 182--85.

29.   Rybian, <u>Podrazdeleniya</u>, 92.

30.   Several pre-battle formations are available to the Soviet
battalion commander, which resemble a line, a wedge, and an
echelon right or left.   Tanks usually lead, with other combat
and support systems placed according to the commander's order.
The interval between units depends on the terrain and enemy
situation, but always supports the rapid deployment of the unit
into a combat formation.   See N. V. Ogarkov, ed., <u>Sovetskaya
voyennaya entsiklopediya</u> [Soviet military encyclopedia], (Moscow:
Voenizdat, 1978), 6:501, sv (<u>Predboyevoy poryadok</u> [precombat
formation]).

31.   D. A. Dragunskiy, ed., <u>Motostrelkovyy (tankovyy) batal'on v
boyu</u>, 78.

32.   Rybian, <u>Podrazdeleniya</u>, 99.

33.   The number preceeding each source corresponds to the example
number in Table 1.

        1   A. Zheltoukhov, "Nastupayet motostrelkovyy batal'on"
[A motorized rifle battalion attacks], <u>VV</u>, No. 2 (1974), 34-38.

        2   M. Kontseropyatov, "Bez pauz - dnem i noch'yu"
[Without pauses - day and night], <u>VV</u>, No. 3 (1980), 32-35.

        3   A. Rybian, "Pod pokrovom nochi" [Under cover of
night], <u>VV</u>, No. 8 (1982), 19-22.

        4   V. Godun, "Batal'on nastupayet noch'yu" [Battalion
attacks at night], VV, No. 11 (1982), 27-30.

        5   I. Semerikov, and A. Chulanov, "Batal'on nastupayet
noch'yu" [Battalion attacks at night], <u>VV</u>, No. 3 (1983), 19-23.

        6   V. Gordeyev, "Men'she pauz pri perekhode ot dnevnykh
deystviy k nochnym" [Fewer pauses during the transition from day
to night operations], <u>VV</u>, No. 4 (1983), 22-24.

        7   K. Babitskiy, and V. Mel'nichuk, "Atakovat'
predstoyalo noch'yu" [The attack was to be at night], <u>VV</u>, No. 1

(1984), 25-28.

8 B. Serbeyev, "Nastupleniye tankovogo batal'ona noch'yu" [A tank battalion night attack], <u>VV</u>, No. 2 (1985), 25-28.

9 Yu. Struchkov, A. Yatsenko, "Svetovoye obespecheniye boya" [Illumination support of battle], <u>VV</u>, No. 3 (1988), 26--28.

34. For purposes of this discussion, a theoretical vulnerability is an aspect of the night attack which, even if executed according to established norms, in the author's opinion might render the Soviet tactical operation vulnerable to detection, disruption, or defeat.

35. K. Kurenkov, "Sovershenstvovat' nochnuyu podgotovku" [Improve night training], <u>VV</u>, No. 2 (1979), 42.

36. <u>Ibid.</u>

37. N. Dudin, "Tankisty uchatsya strelyat' noch'yu" [Tankers learn how to fire at night], <u>VV</u>, No. 2 (1979), 70-72.

38. L. A. Ryazonov, "Sovershenstvovat' nochnuyu podgotovku" [Improve night training], <u>VV</u>, No. 9 (1980), 22-25.

39. Yu. Kulishov, "Nastupayet motostrelkovyy batal'on" [A motorized rifle battalion attacks], <u>VV</u>, No. 6 (1981), 17-20.

40. General Colonel V. Belikov, "Uchit'sia voevat' noch'iu" [Learn to fight at night], <u>VV</u>, No. 8 (1982), 3.

41. V. Luk'yanchuk, "Uchim kursantov deystvovat' noch'yu" [We teach officer cadets to operate at night], <u>VV</u>, No. 1 (1983), 47-48.

42. "Nochnaya podgotovka voysk" [Night training of troops], <u>VV</u>, No. 8 (1983), 2-5.

43. S. Yermolenko, "Motostrelki nastupayut noch'yu" [Motorized rifle units attack at night], <u>VV</u>, No. 4 (1987), 24-27.

44. Because these training weaknesses are exposed with such regularity, one should not presume that Soviet units can do nothing well at night. Rather, the Soviet habit of <u>samokritika</u> [self-criticism] in the military press serves to highlight those aspects of military training which are most difficult to execute well, and thus need continual training emphasis.

45.45. V. A. Merimskiy, <u>Takticheskaya podgotovka motostrelkovykh i tankovykh podrazdeleniy</u> [Tactical training of motorized rifle and tank subunits], (Moscow: Voyenizdat, 1984), 43. The exact same material is reprinted in the 1987 edition of

Merimskiy's book, 43--45.

46. <u>Ibid.</u>, squad in the defense, 75-81 (1984 edition), 90-97 (1987 edition); MRB exercise, 196-216 (1984 edition), 291-314 (1987 edition).

47. G. Krivosheyev, "Polevoy vyuchke - vysokoye kachestvo" [High quality for field training], <u>VV</u>, No. 6 (1985), 20-23.

48. Yu. Groshev, "Nepreryvnyye nochnyye zanyatiya" [Continuous night exercises], <u>VV</u>, No. 4 (1988), 33-37.

49. Specific mentions were made of this use of air defense weapons in examples 4 and 8.